To Know This Place:

The Black Oak Savanna/ Tallgrass Prairie of Alderville First Nation

2nd edition

In eternal gratitude, this book is dedicated to the First Peoples who knew this place. Their stewardship has provided a natural heritage for us that will continue to be a living legacy for generations to come.

Sweetgrass Studios
Alderville First Nation

A Sweetgrass Studios Publication

114 Vimy Ridge Road
R.R.#2, Roseneath, Ontario K0K 2X0 Canada

Copyright © 2005

All rights reserved
No part of this book may be reproduced or transmitted in any form by any means without permission from the publisher.

Sweetgrass Studios trade paperback ISBN 0-9730420-3-6

The Sweetgrass Studios World Wide Web site address is
www.rickbeaver.com

Library and Archives Canada has catalogued this edition as follows:

Library and Archives Canada Cataloguing In Publication

Written by Ruth Clarke
ISBN 0-9730420-3-6

Clarke, Ruth, 1950-
To Know This Place: The Black Oak Savanna/Tallgrass Prairie of Alderville First Nation.—2nd ed.

1. Prairie ecology—Ontario—Alderville region. 2 Prairie flora—Ontario—Alderville region. I. Title.

QH77.C3C52 2005   577.4'4'0971357
C2005-907364-0

Front Cover Photo: Perennial Blue Lupines, courtesy Amanda Newell
Back Cover Photo: A Summer Hike on the Savanna, courtesy Amanda Newell

Table of Contents

Reflections on the Savanna	6
Canada's Easternmost Prairie	7
Introduction: Our Work Continues	8
Fire: The Competitive Edge...	10
Habitat Stewardship	13
Species at Risk	14
Aerial Photo of the Site	16
The Savanna Trail System	17
Black Oaks and Aspens—1	18
The Bowl—2	19
Aspect—3	21
The Hog's Back—4	22
The Foot of the Slope—5	23
Bearberry—6	25
The Draw—7	26
Mutant Red Pine—8	27
Field Thistle—9	28
Sharp-leaved Goldenrod—10	29
Encroaching Forest—11	30
Future Plans—12	31
Old Field Reclamation—13	32
Woodland Trail—14	34
Our Feathered Friends	36
Flowering Plants in May	37
Flowering Plants in June	38
Flowering Plants in July	40
Flowering Plants in August	41
Butterfly Species in Various Habitats	43
Butterflies Seen on the Site	44
Monarch Watch	48
Our Visitors	49
Acknowledgements	50
Where is the Site?	51
Resources	52
Quick Native Plant Species Reference Chart	53

Reflections on the Alderville Savanna

by Rick Beaver

    I recall my early walks in the savanna, motivated by a responsibility to inform others about its special status in the scheme of things and to promote its preservation. Yes, there was a need to document particulars of uncommon plants, animals and relationships among them, but as I became more familiar with the sensory beauty of the place, I realized another of its merits as a wellspring of inspiration, satisfaction and wonder. It remains very much that way with me.

    The lilt of a resting Henry's Elfin butterfly's wings still fascinates me whenever I am lucky enough to encounter one on a sunny spring day on the main trail. This, I had never seen before in my life. Similarly, the diminutive magenta blooms of a Sand Milkwort bobbling in a warm summer wind on the edge of a Little Bluestem patch can inspire new wonder in me.

    Each one of these encounters raises the question, "How and why do these things come to be?" Since those early days, many others have discovered the allure of the place and continue to help answer that question. We all seem to be struck with the notion that whatever our conclusions, the process of our involvement here is a privilege, an opportunity unique in the whole world. On this point we agree.

    I can say that I am very proud of my community for opening the doorway to this world and that we are sharing it with others. There may not be many turns in the path of life where we can say, "I have never been anywhere quite like this before." It is my hope that all who have or will share in this experience take great delight in the journey.

Black Oaks
& Little Bluestem

Canada's Easternmost Prairie

Tallgrass prairie and savanna are among the most endangered plant communities in Ontario. Historically, native grasslands covered more than 800 square kilometers of southern Ontario, stretching from Walpole Island on Lake St. Clair, to the Rice Lake Plains of east central Ontario. With the advent of European settlement, most of these habitats were converted to agriculture and other land uses. Currently, less than 1/10 of one percent of this historical area remains.

The Alderville Black Oak Savanna, located on the Alderville First Nation in Northumberland County, represents the largest single remaining block of tallgrass prairie and savanna in east central Ontario. Originally part of a more extensive grassland complex on the Rice Lake Plains, the site is one of the few remaining examples of Canada's easternmost prairie. — R.B.

Indigo Bunting
(Passerina cyanea)

Eastern Meadowlark
(Sturnella magna)

Golden-winged Warbler
(Vermivora chrysoptera)

Janean Sharkey stands amid 12 1X1 metre plots that she staked out to study cover crop and seed density

2005 Summer assistant Travis Turner monitors Perennial Blue Lupine plots

Introduction: Our Work Continues...

It has been two very short but enormous years since **TO KNOW THIS PLACE: The Black Oak Savanna/Tallgrass Prairie of Alderville First Nation** was first published in 2003. Since that time, in addition to our on-going stewardship of the site, we have added a new trail, implemented new studies and projects, and collected data. We have identified more species of flora and fauna, some rare and many uncommon— corroborating what many academics in the science world have nicknamed the Alderville Savanna: "The Jewel" (of the savanna/tallgrass prairies in Ontario). They refer to the uniqueness of the place, an environment that is always revealing itself to us.

The Savanna has, and continues to receive attention: countless groups from all over the world visit. Young children take their daycare outings here. Students from elementary and secondary schools come here to learn about what grows and lives here. Many college and university students have chosen the Savanna as a study site for science projects, but also as a consideration in ecotourism, ecology management and even firefighting. The Savanna itself seems to appreciate the attention and care it is receiving and has been responding floriferously!

Savanna habitat has grown to include a reclaimed field that Rick Beaver and Dave Mowat planted with seed in autumn 2002. The plants are now mature, flowering and proliferating. Perennial Blue Lupines which volunteers planted in spring 2002 and each year since then, have bloomed and spread, encouraging news for the Karner Blue Recovery Programme. In 2003, a new fly in the Psila genera (*Diptera:Psilidae)* was discovered by a Master of Science candidate from the University of Guelph when he explored the Savanna.

In Spring 2004, due to health problems, Rick Beaver retired as Natural Heritage Coordinator and was succeeded by Amanda Newell. Rick's passion for the Savanna endures and he continues to remain involved, conducting occasional tours and offering advice when requested.

Since the beginning, summer students have brought their own valuable expertise to the site. We now have identified dragonflies, moths, and other insects. We have an impressive bird list that includes a number of species classified by the Ontario Ministry of Natural Resources as either endangered or of special concern because of their dwindling populations in the province. For the past two years, monarch butterflies on the Savanna have been tagged to monitor their migration to their wintering home in central Mexico. Boxes for Eastern Bluebirds and Pileated Woodpeckers have been erected and are cleaned, maintained and monitored each year. To use an old cliché, the Savanna is a going concern.

For this edition, we have added 25 pages, expanding our native plant species list to 15 pages. We have added the Woodland Trail to our Interpretive Points. You will find four pages devoted to butterflies and a page of birds in addition to several placed throughout the book. Students from Roseneath Centennial School participated in a writing contest and three of their quotes appear on the back cover.

Now it's time to reflect on what we've learned on the journey of coming "to know this place." Time to share what we've learned about this jewel. We hope you enjoy your journey exploring our second edition of **To Know This Place: The Black Oak Savanna/Tallgrass Prairie of Alderville First Nation.**

Amanda Newell
Natural Heritage
Coordinator

Fire—The Competitive Edge for Savanna Plants

Savanna plants are warm-season species, and they constantly compete with Eurasian cool-season plants which arrived with settlers. For native plants growing in savannas and prairies, fire is like a rebirth. It extends the growing season for them as much as a month, and shortens the season for invasive Eurasian weeds. Fire cleans away the scrub and brush from beneath the oak trees, allowing more light to penetrate, encouraging native plants to grow.

After a fire, the soil is black and attracts sunlight which warms the soil that would otherwise be matted with leaf litter, acting like a wet blanket. The native plant buds are waiting beneath the soil, and with the eradication of competition, these plants virtually pop through the soil after a burn and following a rain.

The Eurasian invasive plants came from cool meadows, and since they normally go dormant in the heat of mid-summer, what better way to totally confuse them from the outset in early spring? The resultant warm soil that occurs after a fire stops these alien plants' roots from growing, giving a competitive advantage to the native plants.

Aside from accidental lightning strikes that combust, ignite and burn trees and grasses, fires have been set by Native populations for thousands of years for several reasons: to improve game habitat, to increase nut and berry production on trees and shrubs, and to clear passages for easier travelling. Burning dead grass in the spring always results in a fast transition to green. Since its designation as a Natural Heritage site, we have prepared an annual burn management plan that we present to Chief and Council for their approval, and to the Roseneath Fire (and Rescue) Department, and with their close scrutiny, Savanna management and volunteers conduct prescribed burns on specific portions of the Savanna.

In addition to prescribed burns, the occasional wildfire is set— either on purpose or by an errant match or cigarette butt from a passing motorist. On Easter weekend in 2004, a week before the planned burn, a wildfire occurred. It burned exactly what had been planned. Serendipitous arson? In 2005, other wildfires burned the majority of the savanna, perilously close to adjacent homes. Not only do these kinds of fires threaten human safety, but there is no place for wildlife to escape from the fire. Each year, our plan is to burn only one third of the site so that the remaining land is conserved for wildlife.

Little Bluestem
(*Andropogon scoparius*) after a burn

    Perhaps another reason the First Peoples burned this land was for the food that emerged after a burn. Fireweed (*Epilobium angustifolium*), a plant native to both North America as well as Europe, appears immediately following a forest fire. Esteemed as an anti-spasmodic, roots and leaves were brewed to treat asthma, whooping cough and hiccups. Not only is it beautiful, as well as having medicinal qualities, the shoots can be eaten and have been likened to the taste and texture of asparagus. The bright magenta flowers and the stalks and leaves of the plant can be eaten as salad greens.

Fireweed (*Epilobium angustifolium*)

Habitat Stewardship

Karner Blue Butterfly (male)
(Lycaeides melissa samuelis)

The Habitat Stewardship Program, an initiative administered by Environment Canada, continues to support our efforts in establishing habitat suitable for the re-introduction of the Karner Blue Butterfly, a species extirpated from Canada in 1990.

This butterfly requires the Perennial Blue Lupine as its host plant. As one of the potential recovery site projects, we have planted approximately 6,000 Perennial Blue Lupines (*Lupinus perennis*) seedlings since 2002, monitoring their success in various sites. And in some areas, they are actually spreading, either by shoots or seed.

Of the five areas where lupines were planted in 2002, The Bowl and the Hog's Back had higher survival rates, so in subsequent years we concentrated our planting in these areas.

In 2003-4, Pak Kin Chan, a Master of Science candidate from York University, was studying the habitat where Karner Blue Butterflies live in the United States and sites in Ontario that are attempting to re-create habitat. He found Alderville to be very unique, and potentially conducive to successful re-introduction.

Whether or not we are successful in loosing Karner Blue Butterflies on the site, the Blue Lupines appear to be at home here.

Flat of Perennial Blue Lupines and mature plants (right) blooming in the Bowl

Species at Risk

Figures fluctuate and differ nationally versus provincially as to the numbers of species that are considered of importance to be listed in Species at Risk assessment for Ontario. Regardless of the status, we know from the list of (extinct, extirpated) endangered, threatened, of special concern, or listed as not of concern—yet—we have several which visit the site. A Great Gray Owl (*Strix nebulosa)* passed through in summer, 2004. The Eastern Hog-nosed Snake has been seen on Alderville First Nation, but not yet on the Savanna.

Other birds which frequent the site include Cooper's Hawk (*Accipiter cooperii)*, Merlin (*Falco columbarius)*, Northern Harrier *(Circus cyaneus*), Northern Goshawk (*Accipiter gentilis)*, Red-tailed Hawk*(Buteo jamaicensis)* Rough-legged Hawk *(Buteo lagopus)* and Sharp-shinned Hawk *(Accipiter striatus)*, not to mention a large population of Indigo Bunting *(Passerina cyanea)*. Osprey (*Pandion haliaetus)* frequently fly over on their way to and from Rice Lake.

Since our first year on the site, we have erected nesting boxes for the Eastern Bluebird (*Sialia sialis)*, a bird that while not endangered, has suffered, with numbers dropping as low as 17% of their population in the 1950s and 1960s from lack of habitat, from pesticide use and from harsh winters. We started with four boxes in the first year and after adding a few each year, we now have 18 nest boxes that are being shared by Eastern Bluebirds and Tree Swallows (*Iridoprocne bicolor)* and the occasional House Wren (*Troglodytes aedon)*.

Tree Swallow
(*Iridoprocne bicolor)*

Each year, we monitor the boxes once a week from May to August, recording stages of development. When chicks fledge, we clean out the nests for our next customers. This year, 83 chicks fledged: 35 Eastern Bluebirds and 48 Tree Swallows.

Eastern Bluebird chicks

In summer 2004, a Pileated Woodpecker (*Dryocopus pileatus*) and a Common Flicker (*Colaptes auratus*) took up residence in the same telephone pole that had been recently erected on the roadside near the entrance to the site. Three Pileated Woodpecker chicks were seen peeking out from the hole—another time with binoculars but no camera! In the 2005 season, four Pileated Woodpecker nesting boxes were built and erected so they don't have to use telephone poles.

Build it and they will come...Bill Newell (right) erects a Pileated Woodpecker nest box in an Eastern White Pine (*Pinus strobus*) tree. Pileated Woodpecker (left) inspects a dead tree. Common Flicker (*Colaptes auratus*) (centre)

An aerial photograph of the site shot in 2005 looking north-east on the Alderville Black Oak Savanna/Tallgrass Prairie. The Hog's Back (3) runs diagonally across the lower portion of the photograph. The woodland (4) meets the savanna (2) which in turn meets the prairie (1). The Bowl (5) is where the lupines thrive, the decommissioned gravel pit (6) is to the north, and to the right, two types of old field (7) some of which has been pasture, some which has been reseeded.

The Savanna Trail System

Following the numbers on the Savanna Trail System, you are entering The Bowl: 2

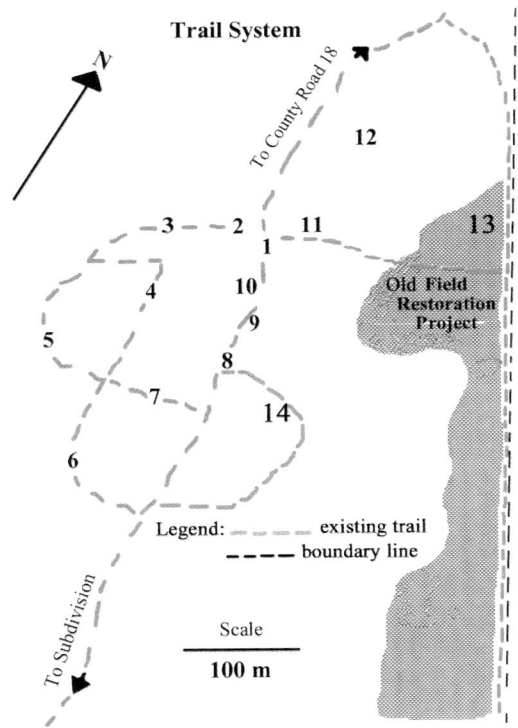

Trail System

Legend: ----- existing trail
        ---- boundary line

Scale 100 m

Key to Interpretive Points on the Savanna Trail System

1: Black Oaks and Aspens
2: The Bowl
3: Aspect
4: The Hog's Back
5: The Foot of the Slope
6: Bearberry
7: The Draw
8: Mutant Red Pine
9: Field Thistle
10: Sharp-leaved Goldenrod
11: Encroaching Forest
12: Future Plans
13: Field Restoration Project
14: Woodland Trail

Black Oaks and Aspens : —1

Black Oak trees were a great resource for Native people. Inner orange bark from the Black Oak, high in tannin content, was harvested in early spring. From dried pulverized bark yellow dye was made. The tannin content was also useful as an antiseptic, tonic and emetic. Infusions were used as a treatment for asthma, as a gargle for colds and hoarseness. Acorns were harvested, washed, boiled and eaten as a dietary supplement.

Three main types of prairie exist in North America: tallgrass, mixed and short-grass prairies, all dependent on particular amounts of rainfall and/ or sharply drained soil. Tallgrass communities with between 10—35% tree cover are called savannas. Alderville's site is called an oak-pine savanna because those trees flourish here.

Tallgrass prairies once comprised one million square kilometers of land in central Canada and the United States. Today, less than one tenth of one percent exists. Our 46.8 hectare remnant (109 acres)—the largest in central Ontario— formed 5000 years ago, after glaciers had melted and a warming trend, suitable soils and fire resulted in a prairie/ savanna habitat known as the Rice Lake Plains.

The Bowl: —2

The Bowl is so named for its shape which has created a microclimate that is cooler and wetter. This area receives more frost but is protected from wind. In this environment plants like Round-headed Bush Clover, Sedge, Dogwood and trees grow, and the Perennial Blue Lupines we planted here like this site best of all the other places we planted.

Because of the practice of burning, many savanna and tallgrass plants growing here are considered uncommon or rare.

Grasses: Big and Little Bluestem and Indian grasses.

Wildflowers: Prairie Lily, Wild Bergamot, Woodland and Thin-leaved Sunflowers, Goldenrods, Prairie Buttercup, Perennial Blue Lupine.

Shrubs: Prairie Willow, New Jersey Tea (near the shade), Serviceberries, Chokecherry, Buffalo Berry.

Trees: White Birch, Aspen Poplar, Red Pine, Black Oak and Black Cherry.

Red Pines & Black Oaks

Roundheaded Bush Clover
(*Lespedeza capitata*)

Pointed-leaf Tick Trefoil
(*Desmodium glutinosum*)

Seeds of the Evening Primrose (*Oeonothera biennis*) are rich in an essential fatty acid used to treat alcoholism, hangover, high blood pressure, hyperactivity, menopause, PMS and weight control. Roots can be eaten as a vegetable, shoots as a salad

Wild Bergamot (*Monarda fistulosa*) is a member of the mint family and is very fragrant. Used for tea (like the flavour of Earl Grey tea) & for flavouring meat

Prairie Lilies *(Lilium philadelphicum)* grow in abundance on the Savanna

New Jersey Tea *(Ceanothus americana)*

Woodland Sunflower var. *(Helianthus divaricatus)*

Indian Grass *(Sorghastrum nutans)* in flower

Little Bluestem *(Schizachyrium scoparium)*

Aspect: —3

Here on the top or north end of the Hog's Back, the soil is acidic, there is sharp drainage and the site is very exposed. Elements of both prairie and savanna exist.

Plants and their growth are affected by soil quality and drainage, exposure to wind, and the amount of moisture they receive. For example, the Early Saxifrage, Prairie Buttercup and Robin Plantain all grow low to the ground, out of the wind and receive moisture from early snow melt. They flower early before the onset of summer heat and drought. The very crusty lichens growing on the ground are also drought resistant.

Robin Plantain was once harvested for medicine. A decoction of the plant and flowers was used to treat tuberculosis, epilepsy and colds.

Prairie Buttercup *(Ranunculus rhomboideus)*

Robin Plantain *(E. pulchellus)*

Upright Bindweed
(Convolvulus spithamaeus)

Early Saxifrage *(Saxifraga virginiensis)*

The Hog's Back: —4

This area of the savanna is a glacial feature and is named Hog's Back for its height and shape, which resemble the animal. This is the driest and windiest part of the site and it is dissected by draws: small channels carved out by glacial water action.

Plants that flourish here are Bearberry (also named Hog Cranberry), Bicknell's Frostweed, New Jersey Tea (*Ceanothus americanus,* also called Red Root, used as a wash for skin problems, as a gargle for throat and mouth sores), Running Serviceberry, Upland Willow, Prairie Buttercup and Prairie Lily (whose bulbs were a Native food).

From this location, to the south and to the north, you can see The Draws. On either side of the Hog's Back are damper areas called the feet of the slope, where vegetation changes. Clearing or restoration must be ongoing to ensure that this part of the trail remains free of spreading undergrowth.

Prairie Brome Grass
(Bromus kalmii)
& Thimbleweed
(Anemone cylindrica)

Field Pussytoes
(Antennaria neglecta)

Sand Milkwort
(Polygala polygama)

The Foot of the Slope: —5

Moisture and nutrients accumulate here, sheltered from the effects of the winds higher up on the Hog's Back. Animals use this edge between prairie and savanna as a preferred trail.

In this environment, White Pine and Chokecherry trees and Gray Dogwood shrubs thrive. Wildflowers found here are Bottle Gentian, Black-eyed Susan, whose plant and roots have been used to treat intestinal worms, snakebites and ear aches, and Wild Bergamot (called Oswego Tea) used for tea and for meat seasoning.

Gray Dogwood
(Cornus racemosa)

Fringed Polygala (*polygala paucifolia*) or Gaywings grow on the eastern side of the Hog's Back

Fall foliage of Low Sweet Blueberry
(Vaccinium angustifolium)

Fruit of the Carrion Flower *(Smilax herbacea)* Flowers of this plant smell like rotting meat. Flies and beetles are attracted to it, maul the flowers and pollinate the plant

Woodland Sunflowers *(Helianthus divaricatus)* & Wild Bergamot (*Monarda fistulosa*)

Hairy Beardtongue *(Penstemon hirsutus)* grows on the Foot of the Slope and on the north part of the Hog's Back

Prairie Cinquefoil *(Potentilla arguta)* & Slender Wheat Grass *(Elymus trachycaulus)*

The Twelve Spot Skimmer (*Libellula pulchela)* is one of many species of Dragonflies to patrol the Savanna alone or in flotilla-like groups

Spreading Dogbane *(Apocynum androsaemifolium)*

Bottle Gentian *(Gentiana andrewsii)* grows in wetter soils at the foot of the Hog's Back and the edge of the swamp

Bearberry: —6

Bearberry plant (left) and blossoms (above)

    The growth of Bearberry *(Arctostaphylos uva-ursi)* indicates an environment where there are warm, dry acidic soils and near full exposure to sun. Associates that also like these conditions are Downy Arrow-wood, Upland Willow, Running Serviceberry, Bicknell's Frostweed and New Jersey Tea.

    Bearberry is also known as Hog Cranberry or *Kinnikinnick*. The leathery evergreen leaves from this plant are ingredients that were combined with tobacco for a Native smoking mixture.

    The Ojibway call this plant *minagunj,* which means "berries with spikes"— the spikes are flower stamens that remain and protrude from the berries. Drooping clusters of waxy pink-white flowers bloom in spring, ripening to red berries that last through winter and are very high in protein (also magnesium, potassium and calcium) but taste astringent and are mealy-textured. Leaves were harvested in early fall before the berries ripened and were also used in concoctions to treat diabetes, sprains, diarrhea, dysentery and many other ailments.

The Draw: —7

After the last glaciers melted 12,000 years ago, this corridor was carved through the Hog's Back, leaving natural travel routes for coyotes, woodchucks, rabbits, deer, raccoons and skunks. The microclimate is slightly wetter and cooler, but sheltered from wind. With more shade and nutrients, plants including Columbine, Lowbush-Blueberry, Fringed Polygala and False Solomon's Seal flourish here. All were important medicinal plants including Seneca Snakeroot, which was used in the treatment of lung disorders.

Early Goldenrod
(Solidago juncea)
& Wild Bergamot
(Monarda fistulosa)

Columbine
(Aquilegia canadensis)

Yellow Warbler
(Dendroica petechia)

False Solomon's Seal
(Smilacina racemosa)

Seneca Snakeroot
(Polygala senega)

Black Snakeroot
(Sanicula marilandica)

Mutant Red Pine *(Pinus resinosa):*—**8**
This extremely rare mutation is called a Witch's Broom. Dwarf conifers propagated from these brooms are highly sought after in the nursery trade for use in landscaping.

Great-spangled Fritillary (*Speyeria cybele*) on Field Thistle

Field Thistle *(Cirsium discolor)*:— 9

This uncommon plant is found in the partial shade of trees at the edge of the savanna. In Native medicine, the plant's roots were ground to a paste and used to treat skin disorders.

Field Thistle provides food for the larvae of the Painted Lady Butterfly (*Vanessa cardui*)

The down from spent field thistle flowers is important to Goldfinches for use in the construction of their nests

Sharp-leaved Goldenrod (above left) blooming in late summer and (above) in early autumn

Sharp-leaved Goldenrod : —10

The Sharp-Leaved or Cut-leaved Goldenrod (*Solidago arguta)* is considered rare in Ontario. Like the Field Thistle, these plants have similar shade-savanna-edge requirements but the Sharp-leaved Goldenrod's larger leaves allow it to grow in shadier (low light) situations. Unlike other species, the flower clusters of this plant are more open. Shady spots are often rich habitat for many birds.

Eastern Phoebe
(Sayornis phoebe)

Least Flycatcher
(Empidonax minimus)

Northern Oriole
(*Icterus galbula*)

Encroaching Forest: —11

Fire is the guardian of the tallgrass prairie; it eliminates dead grasses and leaf litter which decays slowly. Without fire, the dead ground cover would discourage new prairie growth after a few years. Fire hinders the process of natural succession in which Aspen Poplar, White Birch, Black Cherry and Pine trees start to grow out into the openings. The frequency, timing and intensity of fires are factors that affect growth and diversity on tallgrass prairies. Fire also changes the nitrogen composition of the soil and the black ash attracts warmth from the sun, promoting rapid spring plant growth.

White-tailed Deer
(Odocoileus virginianus)

Ruffed Grouse *(Bonasa umbellus)* nests in the encroaching forest and woodland

New growth of Wild Bergamot appears after a fire

Blue Jay
(Cyanocitta cristata)

Future Plans —12

To date, the focus of our work has been on managing the 31 hectares of land that Alderville First Nation has designated as a Natural Heritage, and the several hectares of prairie, savanna, woodland and old field that are directly adjacent this protected area. The continued management and restoration of this core habitat is vital, but it is also important to look beyond these surveyed boundaries. Many private landowners in the Alderville area are privileged to have growing on their properties some of the rarest habitat in North America. With their cooperation, project leaders aim to assess this habitat and inventory plants that grow there, at the same time, passing along information about the ecology and stewardship of their land. Our goal is to create a network of secure tallgrass habitat throughout Alderville on both private and community lands in order to protect all the special species that depend on it.

Prairie ferns and wildflowers

House Wren singing
(Troglodytes aedon)

Cecropia moths mating

Yellow-rumped Warbler
(Dendroica coronata)

Old Field Reclamation——13

In 2002, Alderville First Nation discontinued leasing-out a 15-acre field that had been rented to a farmer for agricultural purposes, and included it in their designation as a natural heritage area. That year, one hectare (2.5 acres) was seeded with prairie species, with the plan of eventually restoring the entire field into healthy grassland habitat. Over the years, the farmer had rotated the crops he planted: two years of grain, two years of corn, then two to three years of forage, a sensible practice that doesn't deplete nutrients from the soil.

Eastern Kingbird
(Tyrannus tyrannus)

Bastard Toadflax
(Comandra umbellata)

Eastern Bluebird
(Sialia sialis)

Hoary Vervain
(Verbena stricta)

The year previous to our reclamation, the field had been planted in oats, a crop that is often used as a cover crop for nursing native plants when they are getting established. Perfect for our purposes. That fall, the reclamation programme began with Rick Beaver and Dave Mowat hand seeding more than 20 kilograms (8 kg per acre) of 28 species of grass and forb seeds: 5 grasses and 23 forbs, collected (also by hand) from the other habitats on the Alderville site. Vermiculite and perlite were mixed with the seed to give better distribution when hand broadcasting.

Later, the seed was harrowed into the soil, it snowed, and Mother Nature took care of the rest until spring.

In June, 2003, the field was mown with the mower blade set 20 centimeters high, to knock down weeds but leave the low prairie plant seedlings untouched and to give them more sunlight. The field was mowed again in August, to knock down invasive weeds before they dropped seed. Of the 28 species planted, 12 were identified growing in the 2003 season. All were forbs: Thimbleweed (*Anemone cylindrica*), Common Milkweed (*Asclepias syriaca*), Butterfly-weed (*Asclepias tuberosa*), Heath Aster (*Aster ericoides*), New England Aster *(Aster novae-angliae*), Azure Aster (*Aster oolentangiensis*), Field Thistle *(Cirsium discolor*), Woodland Sunflower (*Helianthus divaricatus*), Wild Bergamot (*Monarda fistulosa*), Evening Primrose (*Oenothera biennis*), Tall Cinquefoil, (*Potentilla arguta*) and Early Goldenrod (*Solidago juncea*)— but we found no recognizable grasses growing. Other specimens were growing there but we couldn't identify them in their seedling state.

In spring, 2004, we conducted a controlled burn on the field and later re-seeded it with almost 15 kg of seed collected on the savanna the year previous. In the 2004 growing season, we were able to identify 36 native species present in our reclaimed field—26 of which were blooming, and therefore going to set seed.

This year, 2005, we saw the same 36 species of native plants, and have also identified 27 species of Eurasian invaders. We're still winning, and with diligent maintenance, we can reduce the weeds and encourage the native plants to take over.

The Old Field in bloom with Goldenrod, grasses and various species of Asters

Woodland Trail—14

Extensive tree and shrub layers define the woodland. While clearings exist in the woodland, overhead canopy generally ranges from 35 to 60 percent. Soils here are mostly sandy and well-drained. A mixed woodland consisting of Trembling Aspen, White Pine, White Birch, Red Oak and White Oak dominates the tree layer, and extensive stands of Round-leaved Dogwood and Poison Ivy are the main characters in the shrub layer. Our original plan was to cut a more extensive trail through the woods, but the Poison Ivy had the final word. The woodlands here have all endured past periodic burns but their location adjacent current housing locations has resulted in the recent suppression of fire, thus allowing this habitat to become well-established.

Saw-whet Owl
(Aegolius acadicus)

Cedar Waxwing
(Bombycilla cedrorum)

Eastern Towhee
(Pipilo erythrothalmus)

American Woodcock
(Philohela minor)

White Baneberry (*Actaea pachpoda*) flower (left) and berries (above)

White Trillium (*Trillium grandiflorum*)

Bellwort (*Uvularia*)

American Toad (*Bufo americanus*)

Grey Treefrog (*Hyla versicolor*)

Wood Betony (*Pedicularis canadensis*)

Downy Yellow Violet (*Viola pubescens*)

Our Feathered Friends

The Black-throated Green Warbler (*Dendroica virens*) is seen in woodland areas

Downy Woodpecker (*Picoides pubescens*) frequents the woodland

Eastern Phoebe (*Sayornis phoebe*) is seen in mixed habitats

Warbling Vireo *(Vireo glivus)* is seen on the savanna and in the woodland

Gray Catbird *(Dumetella carolinensis)* is most often seen in thickets

Magnolia Warbler (*Dendroica magnolia*) likes the open savanna

Solitary Vireo (*Vireo solitarus*) frequents the woodland

36

Flowering Plants in May

Blue-eyed Grass
(Sisyrinchium montanum)

Wild Geranium
(Geranium maculatum)

Wild Geraniums are also known as Cranesbills, which are what their fruit resembles. The species, *Geranium maculatum,* is said to have high tannin content, with astringent qualities. Used to treat diarrhea both in humans and animals. Also used for treating sores, external bleeding, dysentery and as a gargle.

Wild Strawberry
(Fragaria virginiana)

Prairie Willow catkins
(Salix humilus)

White Trilliums
(*Trillium grandiflorum,*)

Flowering Plants in June

Downy Arrow-wood *(Viburnum rafinesquianum)*

Butterfly-weed (*Asclepias tuberosa*)

Butterfly-weed, also known as Pleurisy Root, was valuable for all chest complaints. An infusion made from the roots was used as an expectorant and anti-inflammatory to treat pleurisy, whooping cough and pneumonia.

Flowering Plants in June

Blue-Eyed Grass
(*Sisyrinchium montanum*)

Smooth Wild Rose (*Rosa blanda*)

Northern Bush Honeysuckle *(Diervilla lonicera)*

Flowering Plants in July

Bicknell's Frostweed
(Helianthemum bicknellii)

Black-eyed Susan *(Rudbeckia hirta)*

Fireweed *(Epilobium angustifolium)* was used as an anti-spasmodic. Infusions made from roots and leaves were effective for treating asthma, whooping cough and hiccups. Their young shoots were eaten like asparagus and the pith from inside the stems was used in soups.

Thimbleweed *(Anemone cylindrica)*. Autumn seeds remain attached to the plant as a cottony "flag" into winter. One of the Savanna's most endearing plants with multiple personalities.

Flowering Plants in August

Thin-leaved Sunflower *(Helianthus decapitalus)*

Sharp-leaved Goldenrod *(Solidago arguta)*

Showy Tick Trefoil *(Desmodium canadense)*

Flowering Plants in August

Calico Aster
(*Aster lateriflorus*)

Large-leaved Aster
(*Aster macrophylla*)

The First Peoples used the roots of the New England Aster to treat fever, catarrh and pain. The Iroquois were said to use it as "a love medicine and as a smudging herb to revive the unconscious."

New England Aster
(*Aster novae-angliae*)

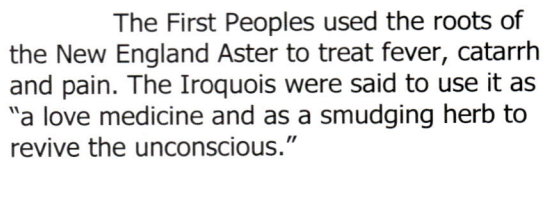

Heath Aster
(*Aster ericoides*)

Azure (Sky Blue) Aster
(*Aster oolentangiensis*)

Butterfly Species Observed in Various Habitats

Savanna	Prairie	Old Field
Hobomok Skipper	Hobomok Skipper	Clouded Sulphur
Arctic Skipper	Indian Skipper	Pink-edged Sulphur
Indian Skipper	Silvery Azure	Common Ringlet
Silver Spotted Skipper	Eastern Tailed Blue	E. Tiger Swallowtail
Crossline Skipper	Dreamy Duskywing	Black Swallowtail
Dun Skipper	Juvenal's Duskywing	Monarch
Spring Azure	Clouded Sulphur	Viceroy
Summer Azure	Pink-Edged Sulphur	American Lady
Silvery Azure	Northern Crescent	
Eastern Tailed Blue	Pearl Crescent	
Dreamy Duskywing	Common Ringlet	
Juvenal's Duskywing	Common Wood Nymph	
Columbine Duskywing	American Lady	
Northern Cloudywing	E. Tiger Swallowtail	
Northern Crescent	Black Swallowtail	
Pearl Crescent	White Admiral	
American Lady	Monarch	
Little Wood Satyr	Viceroy	
Common Wood Nymph	Milbert's Tortoiseshell	
Mourning Cloak	Leonard's Skipper	
E. Tiger Swallowtail	Baltimore Checkerspot	
White Admiral	American Lady	
Elfin (sp.)	Red Admiral	
Banded Hairstreak		
Coral Hairstreak		
Edward's Hairstreak		
Appalachian Brown		
Eyed Brown		
Aphrodite Fritillary		
Great Spangled Fritillary		
Question Mark		
Red Admiral		
American Lady		

Butterflies Seen on the Site

Red Admiral *(Vanessa atalanta)* on Running Serviceberry

Painted Lady *(Vanessa cardui)*

Appalachian Brown *(Satyrodes appalachia)*

Monarch *(Danaus plexippus)*

Eastern-tailed Blue (female)
(*Everes comyntas*)

A puddle of Spring Azures
(*Celastrina ladon*)

Male Black Swallowtail
(*Papilio polyxenes*)

Baltimore Checkerspot
(*Euphydryas phaeton*)

Clouded Sulphur (*Colias philodice*)
on Knapweed

Common Wood Nymph
(*Cerecyonis pegala*)

Coral Hairstreak
(*Satrium titus*)

Little Wood Satyr
(*Megisto cymela*)

Eastern Tiger Swallowtail
(*Papilio glaucus*)

Juvenal's Duskywing
(*Erynnis juvenalis*)

Dreamy Duskywing
(*Erynnis icelus*)

Northern Crescent
(Phyciodes cocyta)

Silver-spotted Skipper
(Epargyreus clarus)

Viceroy
(Leminitis archippus)

White Admiral
(Leminitis arthemis)

Henry's Elfin
(Callophrys hernici)

Monarch Watch

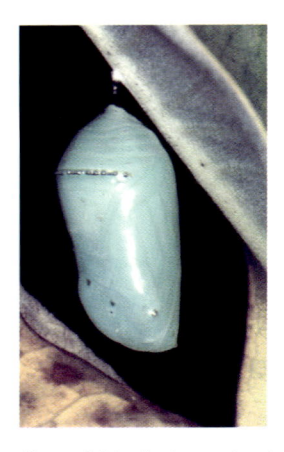

Monarch caterpillar (left) and chrysalis both on Milkweed (*Asclepias syriaca*)

     The plight of the Monarch Butterfly (*Danaus plexippus*) has been of great concern during the past number of years. Their numbers were low in the cool wet 2004 season when we first began tagging, and we were able to tag only four butterflies for Monarch Watch, a data-retrieval programme at the University of Kansas which tracks the migration of the Monarchs.
     The 2005 season was hot and dry: perfect conditions for them. Their breeding season was long, their numbers were high and over the course of the summer we were able to tag 22 butterflies.

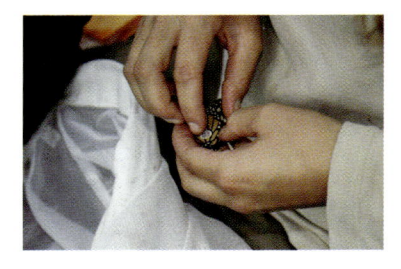

Monarch Butterfly (*Danaus plexippus*) on Heath Aster (*Aster ericoides*) and being tagged

Our Visitors

Rick Beaver (above) shows plants to students from Sir Sandford Fleming College. On the woodland trail (left)

After a 3 km hike on the savanna, these students still appear to have energy to spare

Buffalo wannabes?
A herd of holstein cows visits the savanna

Acknowledgments

Volunteers planting Perennial Blue Lupines in the Bowl

We are grateful to many people and organizations: firstly, to the Government of Canada Habitat Stewardship Program for Species at Risk, whose support has allowed Alderville First Nation to preserve and maintain this site as a natural heritage. We thank the Alderville Community Trust for providing the funds to publish our second edition, and to the photographers who generously provided their shots of the savanna's flora and fauna: Rick Beaver, Bill Crowley, Chris Gooderham, Chris Holmes, Amanda Newell, Bronwyn Salmon, and photographs taken by the late Laszlo (Les) Udvarhelyi. To Maureen Dietrich for her careful editing, to Rick Beaver, for his technical and scientific expertise and his sharp eye for typos.

A very large thank-you to all the committed volunteers who assist us in maintaining the site: the volunteer programme of Nature Conservancy of Canada, Willow Beach Field Naturalists, and to the many individuals who assist us in burns, lupine planting and seed collection. To Jeff Beaver for building Bluebird and Pileated Woodpecker nest boxes and benches. To Bill Newell for erecting the Woodpecker boxes, and for his continuous advice and support. To Ed Heuvel for his on-going support from Halloway Farms. And to Derek Jean for many hours of technical services. Their efforts are invaluable to us—and the savanna.

Canada

Family outing. Kate Hayes of Environment Canada, and her family plant lupines

Where is the Site?

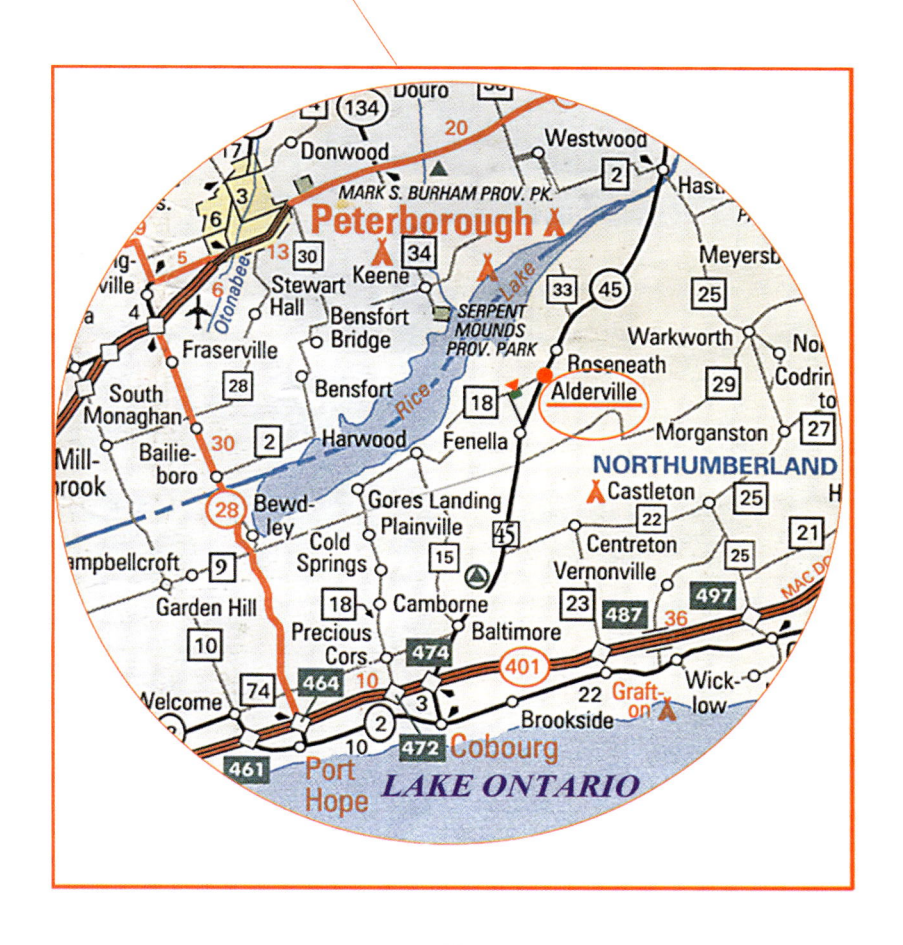

Resources

Halloway Farms Plant Nursery Operated by Ed Heuvel, B.Sc., Halloway Farms supplies a diverse selection of seeds, plugs and plants of native wildflowers and grasses. Halloway Farms provides a complete service of site evaluation, project planning and advice, preparation and maintenance of the site, as well as seeding and planting. Ed offers a broad area of expertise as an ecological consultant, advising in bioengineering and habitat restoration. He also is available for presentations and tours. 55 Halloway Road, R.R.#4 Stirling, ON K0K 3E0 Phone (613) 395-6120 or e-mail: edheuvel@allstream.net

Rice Lake Plains Joint Initiative is comprised of the Nature Conservancy of Canada, County of Northumberland, Ontario Parks, Lower Trent Conservation, Ganaraska Conservation Authority and the Wetland Habitat Fund. The conservation partners envision restoration and protection of sustainable tallgrass prairie and oak savanna habitats on the Rice Lake Plains through co-operative efforts in conservation science, land stewardship, public outreach, and the legal protection of land. To learn how you can protect one of Ontario's most imperiled habitats, visit www.natureconservancy.ca or e-mail todd.farrell@natureconservancy.ca

Wild Ginger Native Plant Nursery and consulting service is operated by Emony Nicholls, B.E.S., offering prairie, wetland and woodland native plant species. Wildflowers, grasses, shrubs and trees particular to the Peterborough area are Wild Ginger's specialty. Emony consults with clients who are interested in ecological restoration and garden design. Visit her at the Peterborough Farmers' Market Saturdays from May to September or via her website www.wildgingernursery.ca, call 705-740-2276 or e-mail wildgingernpn@yahoo.ca.

Willow Beach Field Naturalists is a charitable organization with 200 members who actively participate in volunteer-oriented endeavors that preserve habitat particularly in Northumberland County. WBFN manages Peter's Woods, a provincial nature reserve, for the Ministry of Natural Resources. Members receive The Curlew, a newsletter. WBFN holds monthly meetings and organizes walks and outings. Write to WBFN, P.O. Box 421, Port Hope, ON L1A 3W4

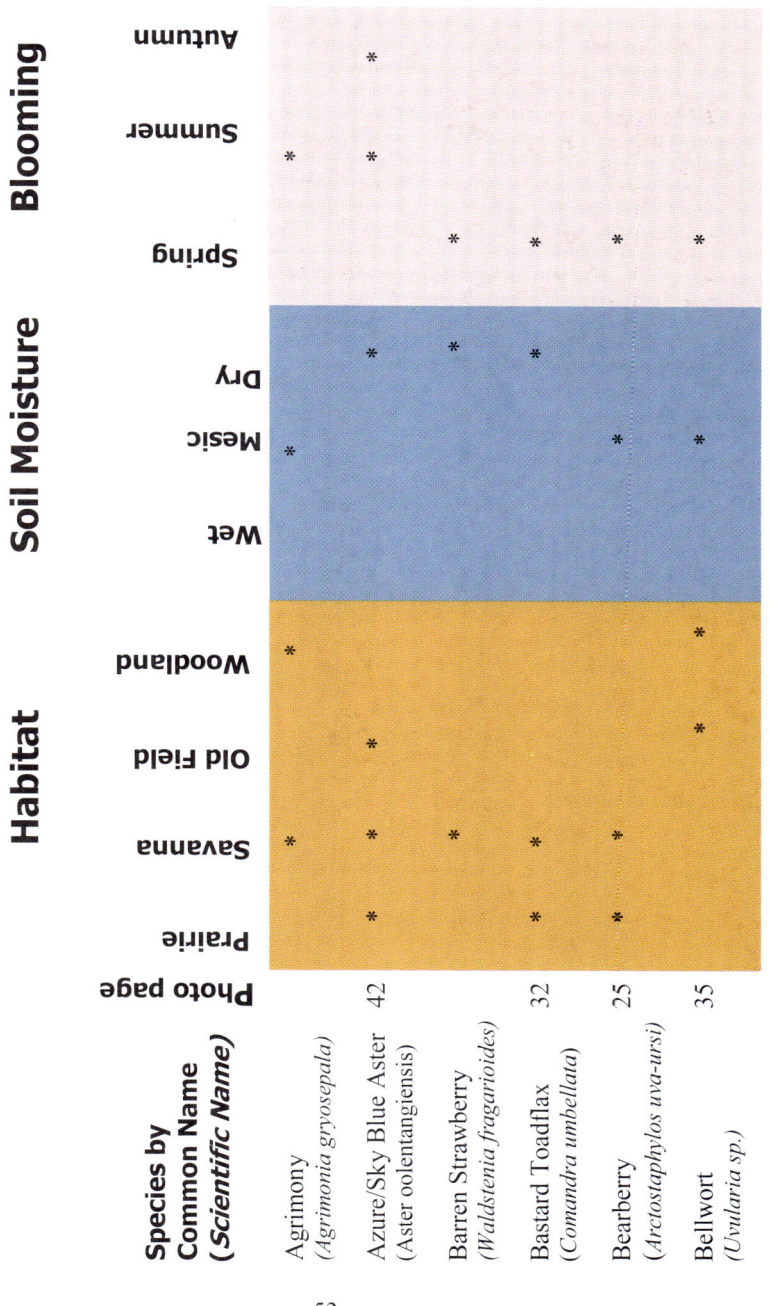

Quick Native Plant Species Reference Chart

Species by Common Name (*Scientific Name*)	Photo page	Habitat				Soil Moisture			Blooming		
		Prairie	Savanna	Old Field	Woodland	Wet	Mesic	Dry	Spring	Summer	Autumn
Agrimony (*Agrimonia gryosepala*)			*		*		*			*	
Azure/Sky Blue Aster (*Aster oolentangiensis*)	42	*	*	*				*		*	*
Barren Strawberry (*Waldstenia fragarioides*)			*					*	*		
Bastard Toadflax (*Comandra umbellata*)	32	*	*					*	*		
Bearberry (*Arctostaphylos uva-ursi*)	25	*	*				*		*		
Bellwort (*Uvularia sp.*)	35			*	*		*		*		

Quick Native Plant Species Reference Chart

Species by Common Name (Scientific Name)	Photo Page	Habitat				Soil Moisture			Blooming		
		Prairie	Savanna	Old Field	Woodland	Wet	Mesic	Dry	Spring	Summer	Autumn
Bicknell's Frostweed/Rockrose (*Helianthemum bicknellii*)	40	*	*				*	*		*	
Big Bluestem (*Andropogon gerardii*)		*	*	*			*	*		*	
Big/large-leaved Aster (*Aster macrophylla*)	42		*		*		*		*		
Black-eyed Susan (*Rudbeckia hirta*)	40	*					*			*	
Black Oak (*Quercus velutina*)	18,19	*	*	*	*		*	*	*		
Black Snakeroot (*Sanicula marilandica*)	26		*	*			*	*		*	
Blue-eyed Grass (*Sisyrinchium montanum*)	39	*	*	*				*	*		
Blue Vervain (*Verbena hastata*)		*		*			*	*		*	

Bottle Brush Grass *(Elymus hystrix)*			
Bottle/Closed Gentian *(Gentiana andrewsii)*	24		
Broadleaved Panic Grass *(Panicum latifolium)*			
Butterfly Weed *(Asclepias tuberosa)*	38		
Canada Fleabane *(Erigeron canadensis)*			
Canada Goldenrod *(Solidago canadensis)*			
Canada Snakeroot *(Sanicula canadensis)*			
Carrion Flower *(Smilax herbacea)*	23		
Columbine *(Aquilegia canadensis)*	26		
Common Cinquefoil *(Potentilla simplex)*			

Quick Native Plant Species Reference Chart

Species by Common Name (*Scientific Name*)	Blooming			Soil Moisture			Habitat				Photo page
	Autumn	Summer	Spring	Dry	Mesic	Wet	Woodland	Old Field	Savanna	Prairie	
Common Fleabane (*Erigeron philadelphicus*)		*	*		*			*		*	
Common Milkweed (*Asclepias syriaca*)		*		*	*			*		*	
Daisy Fleabane (*Erigeron annuus*)		*		*	*			*		*	
Dog Violet (*Viola conspersa*)			*	*			*		*		
Downy Arrow-wood (*Vibernum rafinesquianum*)		*		*			*		*		38
Downy Yellow Violet (*Viola pubescens*)		*		*					*		35
Early Buttercup (*Ranunculus fascicularis*)			*	*					*	*	
Early Goldenrod (*Solidago juncea*)		*		*				*	*	*	26

Early Meadow Rue *(Thalictrum dioicum)*		*	*
Early Saxifrage *(Saxifraga virginiensis)*	21	*	*
Enchanter's Nightshade *(Circaea quadrisculata)*		*	*
Erect /upright Bindweed *(Convulvus spithamaeus)*	21	*	*
Evening Primrose *(Oenothera biennis)*	19	*	*
False Solomon's Seal *(Smilacina racemosa)*	26	*	*
Field Pussytoes *(Antennaria neglecta)*	22	*	*
Field Thistle *(Cirsium discolor)*	28	*	
Fireweed *(Epilobium angustifolium)*	12, 40		*
Fly Honeysuckle *(Lonicera candensis)*		*	*

Quick Native Plant Species Reference Chart

Species by Common Name (*Scientific Name*)	Photo Page	Habitat				Soil Moisture			Blooming		
		Prairie	Savanna	Old Field	Woodland	Wet	Mesic	Dry	Spring	Summer	Autumn
Foamflower (*Tiarella cordifolia*)					*		*		*		
Fragrant Bedstraw (*Galium triflorum*)					*		*		*	*	
Fringed Brome Grass (*Bromus ciliatus*)			*				*	*		*	
Fringed Polygala (*Polygala paucifolia*)	23	*	*					*	*		
Frostweed (*Helianthemum canadense*)		*						*		*	
Gray Goldenrod (*Solidago nemoralis*)		*	*	*				*		*	
Hairy Beardtongue (*Penstemon hirsutus*)	24	*	*					*	*	*	
Hairy Goldenrod (*Solidago hispida*)		*	*	*				*		*	
Hairy Rock Cress (*Arabis hirsuta*)			*		*			*	*		*

Heart-leaved Aster *(Aster cordifolius)*		
Heath Aster *(Aster ericoides)*	16, 42	
Hoary Vervain *(Verbena stricta)*	32	
Hog Peanut *(Amphicarpaea bracteata)*		
Horsetail Fern *(Equisetum hyemale)*		
Indian Grass *(Sorghastrum nutans)*	20	
Lady Fern *(Athyrium felix-femina)*		
Lance-leaved Violet *(Viola lanceolata)*		
Little Bluestem *(Schyzachrium scoparius)*	20	
Long-stalked Panic Grass *(Panicum perlongum)*		

Quick Native Plant Species Reference Chart

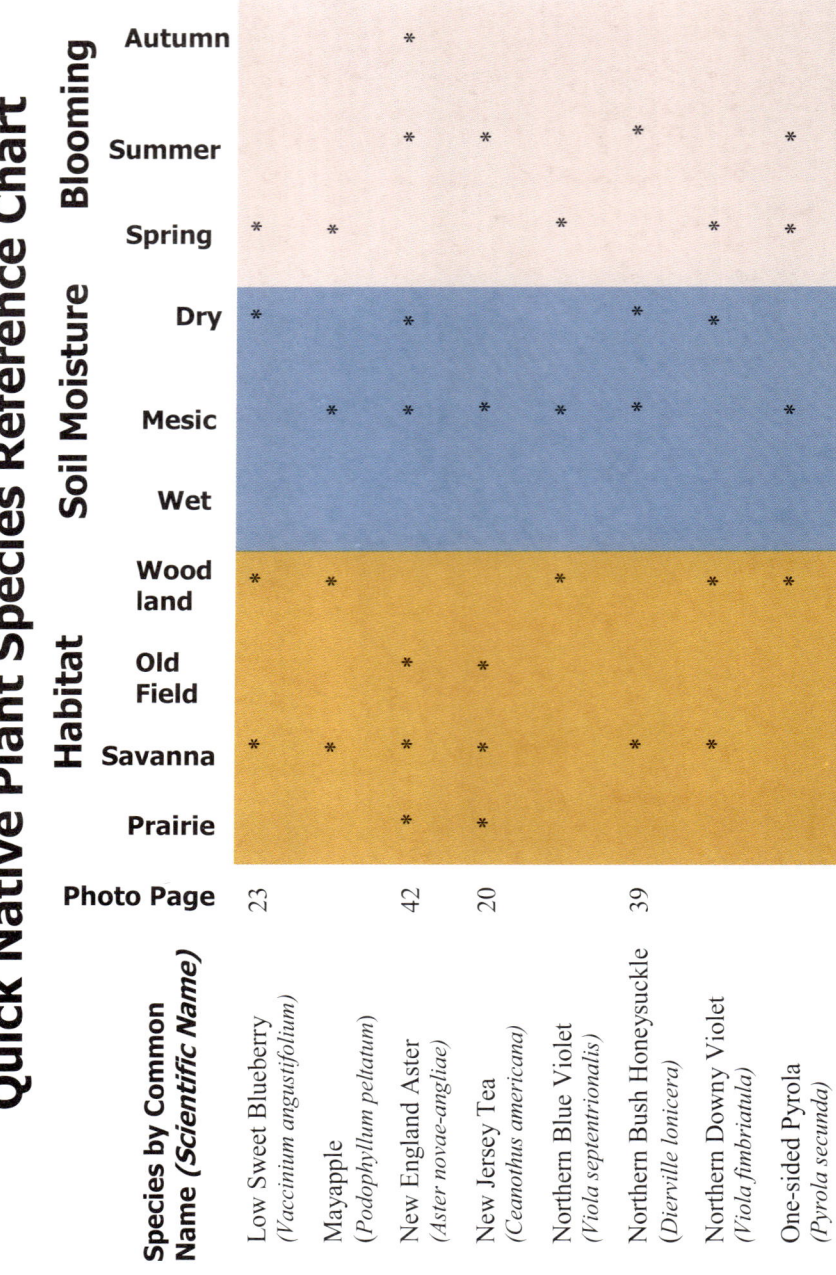

Species by Common Name (*Scientific Name*)	Photo Page	Habitat				Soil Moisture			Blooming		
		Prairie	Savanna	Old Field	Woodland	Wet	Mesic	Dry	Spring	Summer	Autumn
Low Sweet Blueberry (*Vaccinium angustifolium*)	23		*		*			*	*		
Mayapple (*Podophyllum peltatum*)			*		*				*		
New England Aster (*Aster novae-angliae*)	42	*	*	*			*	*			*
New Jersey Tea (*Ceanothus americana*)	20	*	*	*			*			*	
Northern Blue Violet (*Viola septentrionalis*)					*		*		*		
Northern Bush Honeysuckle (*Dierville lonicera*)	39		*				*	*		*	
Northern Downy Violet (*Viola fimbriatula*)			*				*	*	*		
One-sided Pyrola (*Pyrola secunda*)					*		*		*	*	

Pale-leaved Sunflower
(Helianthus strumosus)

Pale St John's Wort
(Hypericum ellipticum)

Pale Vetchling
(Lathyrys ochroleucus)

Pearly Everlasting
(Anapahalis margaritacea)

Pensylvanica Sedge
(Carex pensylvanica)

Perennial Blue Lupine
(Lupinus perennis) Intr.   13, cover

Plantain-leaved Pussytoes
(Antennaria parlinii ssp falax)

Pointed Blue-eyed Grass
(Sisyrinchium augustifolium)

Pointed-leaf Tick Trefoil   19
(Desmodium glutinosum)

Poison Ivy
(Rhus radicans)

Poke Milkweed
(Asclepias exaltata)

Quick Native Plant Species Reference Chart

Species by Common Name *(Scientific Name)*	Photo Page	Habitat				Soil Moisture			Blooming		
		Prairie	Savanna	Old Field	Woodland	Wet	Mesic	Dry	Spring	Summer	Autumn
Poverty Grass/False Heather *(Hudsonia tomentosa)*		*	*					*	*	*	
Poverty Grass *(Danthonia spicata)*		*	*		*			*		*	
Prairie Brome *(Bromus kalmii)*	22	*	*				*	*		*	
Prairie Buttercup *(Ranunculus rhomboideus)*	21	*	*					*	*		
Prairie/Tall Cinquefoil *(Potentilla arguta)*	24						*	*		*	
Prairie/Wood Lily *(Lilium philadelphicum)*	20	*	*					*	*		
Prairie Willow *(Salix humilis)*	37	*	*					*	*		
Purple Oat Grass/False Medic *(Schiachen pupurascens)*			*				*	*	*	*	
Purple-stemmed Aster *(Aster puniceus)*			*		*	*				*	*

Rattlesnake Root/White Lettuce *(Prenanthes alba)*			
Red Baneberry *(Actaea rubra)*			
Red Pine *(Pinus resinosa)*	19		
Richardson's Sedge *(Carex richardsonii)*	21		
Robin Plantain *(Erigeron pulchellus)*			
Rose-twisted Stalk *(Streptopus roseus)*			
Rough Cinquefoil *(Potentilla norvegica)*			
Rough Fleabane *(Erigeron strigosus)*			
Round-headed Bush Clover *(Lespedeza capitata)*	19		
Roundleaf Pyrola *(Pyrola rotundiflora)*			
Round-leaved Dogwood *(Cornus rugosa)*			
Sand Milkwort *(Polygala polygama)*	22		
Seneca Snakeroot *(Polygala senega)*	26		

Quick Native Plant Species Reference Chart

Species by Common Name (*Scientific Name*)	Photo Page	Habitat: Prairie	Savanna	Old Field	Woodland	Soil Moisture: Wet	Mesic	Dry	Blooming: Spring	Summer	Autumn
Sharp-leaved Goldenrod (*Solidago arguta*)	29, 41		*		*					*	
Shinleaf (*Pyrola elliptica*)	41				*		*			*	
Showy Tick Trefoil (*Desmodium canadense*)			*	*			*				*
Siccata Sedge (*Carex siccata*)			*				*		*		
Silverrod (*Solidago biocolor*)			*					*		*	
Sleepy Catchfly (*Silene antirrhina*)		*	*					*	*	*	
Slender Panic Grass (*Panicum xanthophysum*)		*	*				*	*		*	
Slender Wheat Grass (*Elymus trachycaulus*)	24	*	*					*	*	*	
Smaller Pussytoes (*Antennaria neodioca*)		*	*		*		*	*	*		

24

22

Snowberry
(Symphoricarpos albus)

Solomon's Seal
(Polygonatum biflorum)

Spinulose Woodfern
(Dryopteris spinulosa)

Spotted Joe-pye Weed
(Eupatorium maculatum)

Spreading Dogbane
(Apocynum androsaemifolium)

Starflower
(Trientalis borealis)

Star-flowered Solomon's Seal
(Maianthemum stellata)

Starved Panic Grass
(Panicum depauperatum)

Tall Goldenrod
(Solidago altissima)

Tall White Lettuce
(Prenanthes altissima)

Thimbleweed
(Anemone cylindrica)

Thimbleweed
(Anemone virginiana)

Quick Native Plant Species Reference Chart

Species by Common Name *(Scientific Name)*	Photo Page	Habitat				Soil Moisture			Blooming		
		Prairie	Savanna	Old Field	Woodland	Wet	Mesic	Dry	Spring	Summer	Autumn
Thin-leaved Sunflower *(Helianthus decapetalus)*	41		*	*			*	*		*	*
White Avens *(Geum canadense)*			*		*		*			*	
White Baneberry *(Actaea pachypoda)*	35				*		*		*		
White-haired Panic Grass *(Panicum villossissimum)*			*		*		*	*		*	
White Mountain Rice *(Oryzopsis asperifolia)*			*		*	*		*	*		
White Trillium *(Trillium grandiflorum)*	37				*		*		*		
Wild Bergamot *(Monarda fistulosa)*	19, 30	*	*	*			*	*		*	
Wild Geranium *(Geranium maculatum)*	37	*	*		*		*	*	*		